Tropical Rainforests

By Sindy McKay

TREASURE BAY
Family Engagement in Reading

Parent's Introduction

Whether your child is a beginning reader, a reluctant reader, or an eager reader, this book offers a fun and easy way to encourage and help your child in reading.

Developed with reading education specialists, **We Both Read** books invite you and your child to take turns reading aloud. On your turn, you read a left-hand page; then your child reads a right-hand page, which has text written at a specific reading level. The result is a wonderful new reading experience with less frustration and faster reading development!

You may find it helpful to read the entire book aloud yourself the first time, then invite your child to participate the second time. As you read, try to make the story come alive by reading with expression. This will help to model good fluency. It will also be helpful to stop at various points to discuss what you are reading. This will help increase your child's understanding of what is being read.

In some books, a few challenging words are introduced in the parent's text with **bold** lettering. Pointing out and discussing these words can help to build your child's reading vocabulary. If your child is a beginning reader, it may be helpful to run a finger under the text as each of you reads. To help show whose turn it is, a blue dot ● comes before text for you to read, and a red star ★ comes before text for your child to read.

If your child struggles with a word, you can encourage "sounding it out," but keep in mind that not all words can be sounded out. Your child might pick up clues about a word from the picture, other words in the sentence, or any rhyming patterns. If your child struggles with a word for more than five seconds, it is usually best to simply say the word.

Most of all, remember to praise your child's efforts and keep the reading fun. At the end of the book, there is a glossary of words, as well as some questions you can discuss. Rereading this book multiple times may also be helpful for your child.

Try to keep the tips above in mind as you read together, but don't worry about doing everything right. Simply sharing the enjoyment of reading together will increase your child's reading skills and help to start your child off on a lifetime of reading enjoyment!

Tropical Rainforests

A We Both Read Book
Level 1–2
Guided Reading: Level H

With special thanks to Emma Kocina, Biologist at the California Academy of Sciences, and Manuel Luján Anzola, Botanist at the Royal Botanic Gardens, Kew, UK, for their review of the information in this book

To Bonnie and Jeremy —and all who come after them.
— S. M.

Use of photographs provided by iStock and Dreamstime.
Map images of Borneo on page 36 are licensed under a Creative Commons Attribution-ShareAlike 4.0 International License. Source: Contributor and Victim - Indonesia's Role in Global Climate Change with Special Reference to Kalimantan.

Published by
Treasure Bay, Inc.
PO Box 519
Roseville, CA 95661 USA

Printed in Malaysia

Library of Congress Catalog Card Number: 2021942356

ISBN: 978-1-60115-370-8

Visit us online at
WeBothRead.com

PR-5-23

Table of Contents

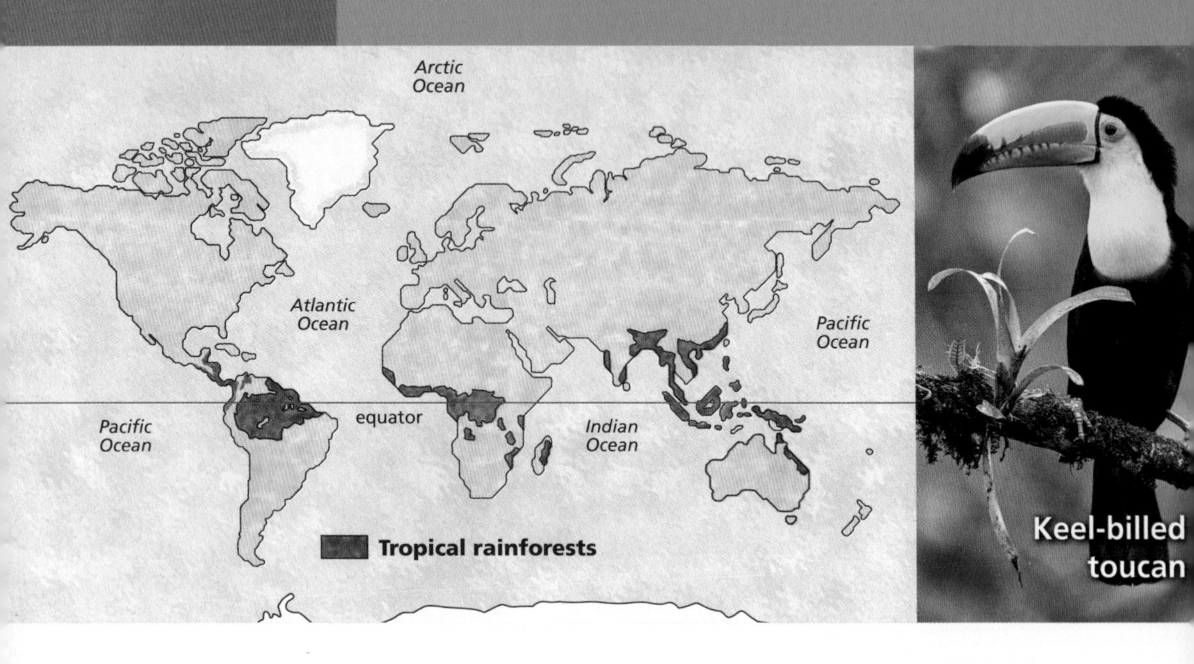

- Located in the warm areas around the world's equator, **tropical rainforests** are one of **Earth's** most diverse ecosystems. An ecosystem is an environment with many kinds of plants and animals that interact and affect the lives of each other. The loss of even one kind of plant or animal can disturb the balance of the ecosystem.

  Tropical rainforests cover only about six percent of the **Earth's** surface, yet these ecosystems are home to over half of **Earth's** plant and animal species.

★ It rains a lot in a **tropical rainforest**. The air is always wet and warm. It feels like summer all year long. Many of the plants and animals here live no place else on **Earth**.

Giant anteaters can eat 30,000 ants in a day.

A rainforest is made up of several layers. The bottom layer is called the **forest floor**. This is on or near the ground. The **forest floor** is filled with ferns, fungi, moss, tree roots, and small bushes. It is home to many types of rainforest cats. You might also run into elephants, anteaters, and very large snakes.

The Rafflesia flower is the largest flower in the world and can be up to three feet wide. It has a really horrible smell that attracts flies.

Forest floor

★ Trees in the rainforest can grow very close to each other. These trees block the sun. That makes much of the **forest floor** dark. It is also hot and wet. There are lots of bugs here. Lizards and frogs like to eat the bugs.

Green forest lizard

Jaguar
Understory

The next layer is called the **understory**. This layer is above the ground and is thick with small trees, strong vines, shrubs, and ferns. Beautiful bright-colored flowers also grow in this layer. These trees and plants make nice homes for a vast assortment of animals, including **jaguars** who spend much of their time in the lower branches of the trees.

Red-eyed tree frog
Lantern bug

Jaguar

★ There are lots of birds, frogs, and snakes in the **understory**. **Jaguars** will come down from the trees at night to hunt. They mostly hunt animals on the forest floor.

The **canopy** layer of the rainforest is made up of tall treetops. Up here there is a lot of sun and food is plentiful. Most of the animals who live in the canopy layer can either fly, jump, or climb.

The tallest trees in the rainforest are found in the **emergent layer**. These are some of the oldest trees, and they have been reaching toward the sun for a long time.

Juvenile green bush viper

Shorea trees can be as tall as a 30-story building.

★ More animals live in the **canopy** layer than any other layer.

The **emergent layer** is a safe place for some animals. The animals who might hunt them can't reach them here.

Bald uakari

CHAPTER 2

WHY WE NEED RAINFORESTS

- Some people call rainforests the "lungs" of the earth. All those trees take in large amounts of carbon dioxide—the bad stuff—and **recycles** it into large amounts of oxygen—the good stuff! Without those trees, our air would be less safe to breathe.

★ The trees in rainforests also help **recycle** the fresh water on Earth and keep it clean. Their roots take up water from the ground. It goes up into their leaves. The water is then sent back into the air as mist and clouds.

Strawberry poison dart frog

Eastern lowland gorilla

Tropical rainforests are home to more than half of our planet's land-based animal species. Because the temperature stays constant and warm, animals don't have to worry about freezing or being too hot. Water and food are plentiful.

Some rainforest animals have a healthy and thriving population. Some species have less than fifty surviving members and are critically endangered. An endangered species is any type of animal or plant that is at risk of disappearing forever.

Emerald tree boa

Scarlet Ibis

Oscar fish

Howler monkey

★ There are also people who live in the rainforests of the world. Many of these people use the forest plants and animals for food. They may also use wood and other parts of the trees to make their homes.

Quinine, which treats malaria, comes from the cinchona tree.

Cacao (kuh-KOW) or cocoa (KOH-koh) tree with cacao pods

A great number of our modern medicines are derived from rainforest plants. These medicines are used to treat malaria, heart disease, cancer, and many other health problems.

The trees in the rainforest also give us some of our favorite foods to eat. **Chocolate** comes from **cacao** (kuh-KOW) trees. Banana and fig trees can be found in almost all tropical rainforests.

Parrot eating a fig

Sangre de drago is often used to heal wounds.

Macaque monkey eating a mango

★ Seeds of the **cacao** (kuh-KOW) fruit are used to make **chocolate**. Rainforest animals love this sweet fruit. They also love bananas, mangoes, and other fruit.

Pygmy marmoset

Aerial view of Amazon rainforest in Brazil

- The Amazon Rainforest is the largest in the world. It contains more than fourteen thousand different kinds of plants and around 400 *billion* trees! That's more trees than there are stars in our galaxy!!

The variety of animals is remarkable. Scientists have found over four hundred species of mammals and suspect there are more than a million different types of insects.

Winding through this rainforest is the Amazon River.

Green vine snake

King vulture

Spectacled caiman

★ Some animals live in the river. Other animals come to the river to find food and drink the water. The river is not always a safe place to be. Some animals with very sharp teeth swim in these waters!

Electric eel

Piranha

Harpy eagle

Yellow banded poison dart frog

The Amazon rainforest is full of fascinating and sometimes frightening animals. The largest mammal in this rainforest is the jaguar, which hunts its prey on the forest floor. The top predator in the canopy is the harpy eagle and in the water is the huge, green anaconda snake. One of the most dangerous to humans is the **poison** dart frog. The skin of some of these frogs holds enough **poison** to kill ten people!

Green anaconda

Margay cat

Scarlet macaw & blue and gold macaw

★ The jaguar is not the only cat in the Amazon rainforest. There are many others. There are also many other kinds of snakes, birds, and frogs. The snake you see on this page is small. But it has **poison** in its a bite that can kill a large animal!

Eyelash viper

Toucan

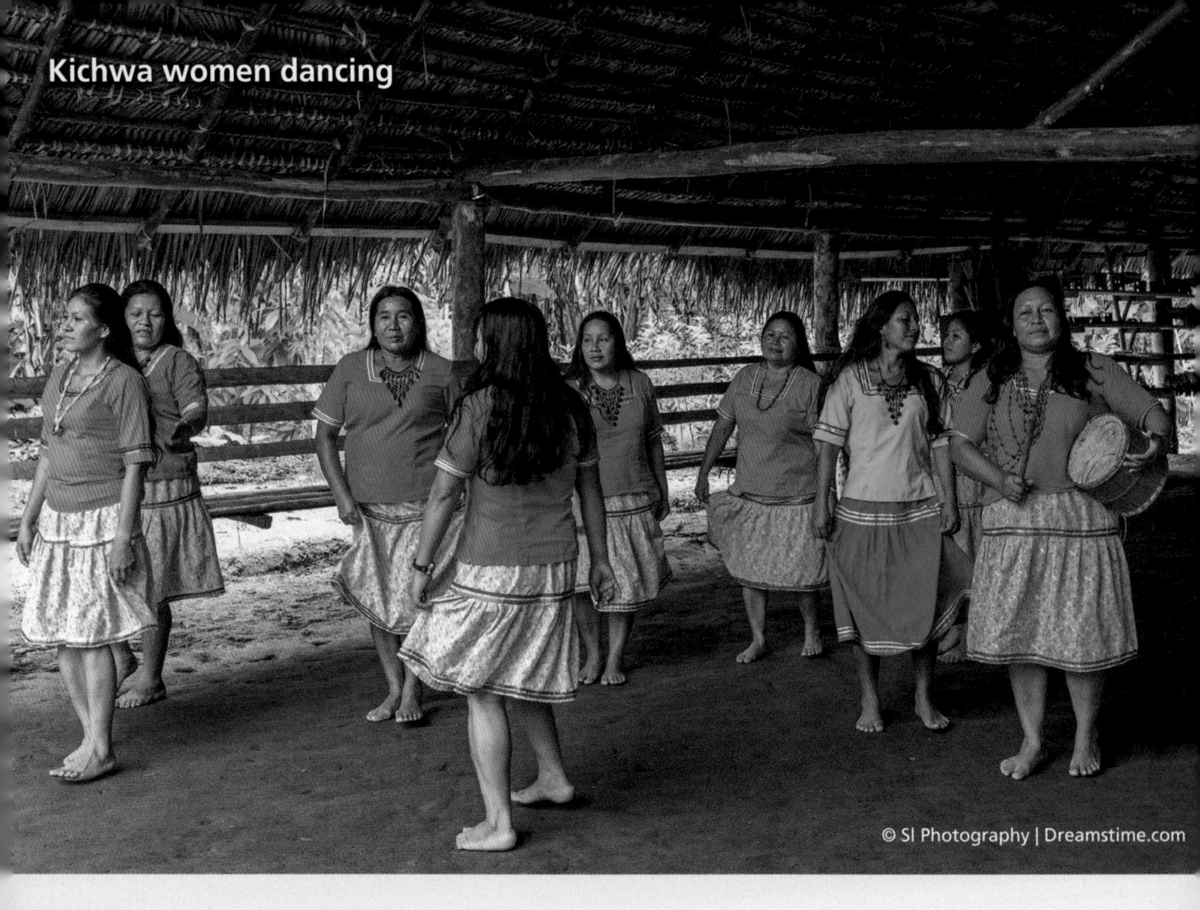

Kichwa women dancing

Not only plants and animals live in the Amazon rainforest. People live here as well. The Kichwas are one of the largest groups of indigenous—or **native**—peoples with a population of over one hundred thousand. They believe that the well-being of a community depends on a strong relationship with nature.

Kichwa girl going to school

Native Brazilian child from Tupi Guarani tribe

★ There are many hundreds of **native** tribes in the Amazon rainforest. The people in many of these tribes have never met anyone outside of their own tribe.

Congo Basin rainforest, Republic of Congo

- The second largest rainforest in the world is in the Congo Basin, which spans six central African countries. The trees in this forest are taller than those in the Amazon, but lower layers are not nearly as thick. This is because large plant-eating animals, such as forest **elephants** and gorillas, feed on the shorter, smaller trees.

Baby gorilla

★ Forest **elephants** are smaller than the African elephants you may see at the zoo. Forest elephants have tusks that point down. African elephants have tusks that point up more.

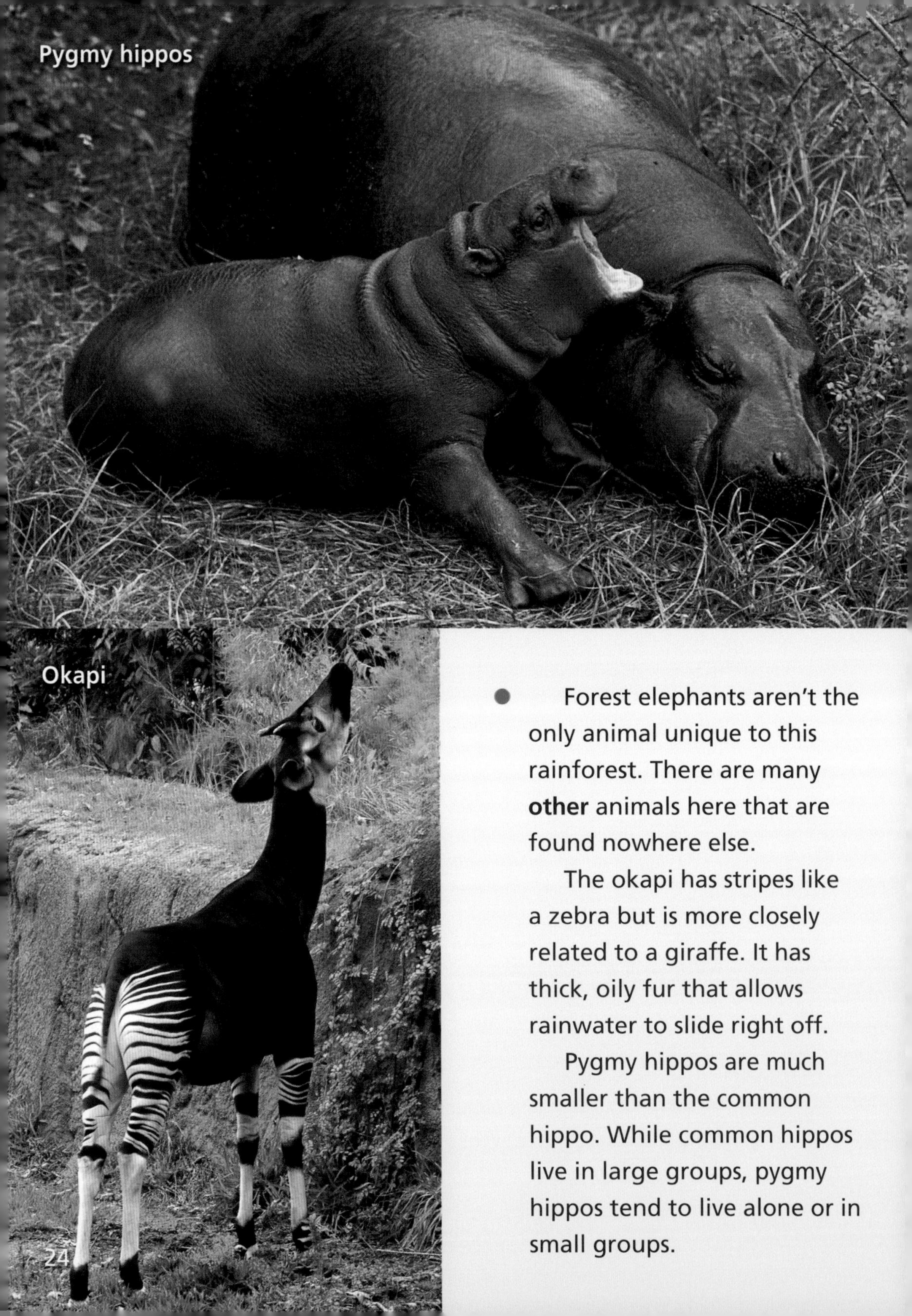

Forest elephants aren't the only animal unique to this rainforest. There are many **other** animals here that are found nowhere else.

The okapi has stripes like a zebra but is more closely related to a giraffe. It has thick, oily fur that allows rainwater to slide right off.

Pygmy hippos are much smaller than the common hippo. While common hippos live in large groups, pygmy hippos tend to live alone or in small groups.

★ Red river hogs swim well and run fast. They eat plants. **Other** animals try to eat them!

The bush viper is a small snake with a big bite. Some people think it looks like a small dragon.

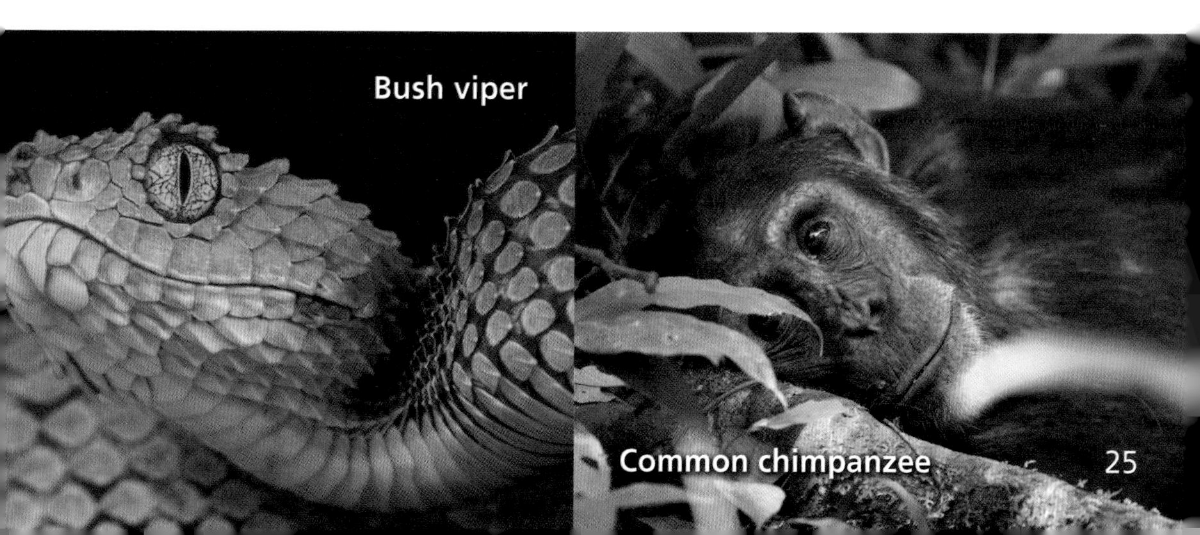

Several remarkable Congo Basin rainforest animal species are in danger of disappearing forever.

There are only about 150 Eastern bongo antelope left in the wild. Bonobo chimps and common chimps are both endangered. The African leopard, eastern lowland gorilla, western lowland gorilla and mountain gorilla are quickly disappearing as well.

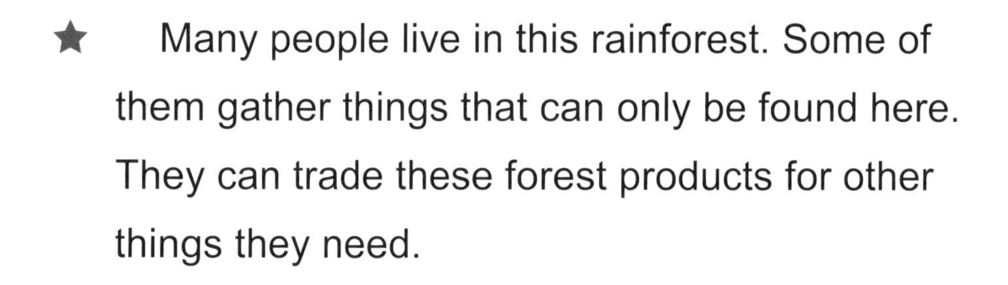

★ Many people live in this rainforest. Some of them gather things that can only be found here. They can trade these forest products for other things they need.

CHAPTER 5 RAINFORESTS OF ASIA

- The oldest rainforests on Earth are in Southeast Asia. Scientists believe that dinosaurs roamed these forests seventy million years ago. There are no more dinosaurs, but there are still some pretty amazing animals that can only be found here.

Orangutans, the world's largest tree dwelling animal, is a great ape that lives in the Asian rainforests of Borneo and Sumatra. Sadly, these beautiful animals are critically endangered.

★ The word “**orangutan**” means “people of the forest.” An orangutan baby rides on its mother’s back. It uses its feet like hands. It can use its feet to grab branches and pick fruit.

Jambu fruit
Durian fruit

Fruit trees are an important resource in Southeast Asian rainforests. Jambu, durian, and strangler fig trees all bear fruit that is eaten by the animals of the forest. Strangler fig trees are perhaps the most important as a large variety of animals eat its fruit.

Orangutans eating durian fruit
Strangler fig tree

Gibbon eating jambu fruit

Coppersmith barbet eating strangler fig fruits

★ Some animals that eat figs are monkeys, gibbons, birds, and bats. Fig trees grow new fruit all year long. The animals always have something to eat.

Flying fox bats

Sumatran tiger
Jambu fruit dove

The rainforests of Southeast Asia have many unique animal species, such as the Sumatran tiger, Jambu fruit dove, Sumatran rhinoceros, Asian elephant, Draco lizard, and Asian tapir. These forests are also home to the greatest number of people who depend on the rainforest to live. Yet these forests are disappearing faster than any others on earth.

Draco flying lizard
Vietnamese moss frog
Sumatran rhino
Asian elephant

★ Many homes in these rainforests are up on stilts. This is to help them stay dry when the big rains come. It also helps keep them safe from animals.

Fire-tailed titi monkey, discovered in 2010

- Many areas within the world's rainforests are still unexplored. New and mysterious creatures are being discovered all the time. There are so many interesting plants that scientists don't have time to investigate them all. Some of these plants may hold the cure to deadly diseases.

Unicorn praying mantis, similar to one discovered in Brazil in 2019

★ There is still a lot more we can learn from the rainforests. There are animals and plants that no human has ever seen! Would you like to explore the rainforests?

Maps showing deforestation since 1950 on the island of Borneo in Southeast Asia. Borneo is the third largest island in the world.

1950

1985

2000

2005

2010

2020

Tropical rainforests

Deforested land

Keeping our tropical rainforests alive and healthy is important, but something called "deforestation" is putting them in grave danger of disappearing. Deforestation means the trees are cut down for lumber or burned down to make space for ranches and farms. An area of rainforest the size of a football field is lost every six seconds!

★ Most rainforest trees have been growing a long, long time. It would take a very long time for them to grow back. Most will never grow back.

Cattle ranching in the Amazon

- Once an area of rainforest has been changed into a cattle ranch or a farm, it is unlikely it will ever return to its natural state. Without this important ecosystem, many plant and animal species will die out. Life-saving plants may never be discovered.

Deforestation in the Amazon

Silverback gorilla

★ We need the rainforests. And the rainforests need us. They need our help to save the plants and animals before they are all gone.

Flock of red-and-green macaws

Jaguar

Dusky leaf monkey

- How can you help to save the rainforests? You have already started by learning about them. Here are some other ways to help.
  1. Ask your parents to buy food—like bananas, coffee, chocolate, and palm oil—that is grown in a way that is safe for the rainforests.
  2. Do a class project to learn even more about these ecosystems. The more you know, the more you can help.
  3. Work with grown-ups to organize a fundraiser to raise money for a rainforest conservation group.

African forest elephant

Red-eyed tree frog

★ Many people are working to help save rainforests. Now you know some ways that you can help too. It's not too late to save the trees and animals of these amazing forests.

Glossary

canopy
the rainforest layer made up of the thick branches and leaves of the taller trees

deforestation
the cutting down of the trees in large areas of the forest

ecosystem
a group of living things that live and interact with each other in a certain environment

emergent layer
the rainforest layer made up of the tallest trees that rise above all the others

forest floor
the dark, damp ground beneath the rainforest trees

understory
the rainforest layer below the canopy made up of bushes, shade-loving plants, and short trees

Questions to Ask after Reading

Add to the benefits of reading this book by discussing answers to these questions. Also consider discussing a few of your own questions.

1. Can you think of any animals that live in the rainforest? Which one is your favorite? Why?

2. What do you think are some reasons that so many animals live in the canopy layer of the rainforests

3. What are some reasons you think people should try to protect the rainforests?

4. Do you think it is important to stop animals from going extinct? Why?

5. Would you like to explore a rainforest someday? What kind of things do you think you might see?

If you liked ***Tropical Rainforests*** here are some other We Both Read® books you are sure to enjoy!

You can see all the We Both Read books that are available at WeBothRead.com.